The Cutty Sark
Her Designer and Builder

H Linton.

PREFACE

The writing of this brief biography of Hercules Linton, surveyor of ships, master shipbuilder, and marine architect—as he described himself at various periods in his life—is an act of family piety. His eldest son, Hercules Anderson Linton, born in Dumbarton in October 1869, was the only one of his large family to marry and have children. My wife is the youngest of the three daughters of Hercules Anderson Linton and a granddaughter of the *Cutty Sark* Linton. There are now great-grandchildren and great-great-grandchildren, some associated in name with the *Cutty Sark* Linton. It is hoped that these and others may be interested in the meagre facts of the interesting career of their forebear, who lived during a time of great change in British shipbuilding, from wood, and wood and iron, to all iron and steel, and from sail to steam. In greater or less degree he no doubt contributed to these developments and with fast steam yachts and iron ships for oil and molasses may have pointed the way for later developments.

I have been able to use family records and papers, and fortunately Hercules Linton left notebooks and many drawings—mainly in connection with the ornamentation on the *Cutty Sark*—as well as a book of newspaper cuttings to show the range of his interests. I am especially grateful to my wife's eldest sister, Mrs Hilda Hunter, for making family correspondence and other papers so readily available, and to my wife's aunt, Miss Ann Linton, the elder of two surviving daughters of Hercules Linton, who in her eighty-ninth year, in spite of certain infirmities of sight and hearing, is still active physically and in excellent memory. My acknowledgment of other help is mentioned in the notes collected at the end of the memoir.

Two debts must be mentioned specially. I am very grateful to the *Cutty Sark* Society for making possible the printing and publication of this biography; and I offer my best thanks to Mr Frank Carr for his enthusiastic interest and continuing encouragement.

ROBERT E. BRETTLE

September 1968

Pencil sketch for name on stern, drawn by Hercules Linton.

ILLUSTRATIONS

The frontispiece is enlarged from the original photograph in the possession of Mrs R. E. Brettle, who also owns, among others, the pencil drawings of the Dumbarton shipyard and of the figurehead. The *Cutty Sark* at anchor in Cork Harbour is from a family oil painting commissioned by Hercules Linton for his father, Alexander.

The *Cutty Sark* in full sail was painted by F. Tudgay for the ship's first owner, Captain 'Jock' Willis, and is now owned by his niece, Mrs M. Kirkaldy-Willis.

Mrs Brettle's old photographs, pencil drawings, old notebooks and the oil painting were loaned to the *Cutty Sark* Society and copies were made for the National Maritime Museum. John Rennie's original sail and rigging plan (not reproduced here) was given by her to the Society in 1968.

The author and publishers are grateful to the Trustees of the National Maritime Museum, and to Mr Frank Carr and the *Cutty Sark* Society, for help in reproducing the illustrations.

CHAPTER 1

Linton is regarded as a locality name. There is a village Linton, now West Linton, in Peeblesshire; Linton a parish in Roxburghshire; and East Linton a small burgh in East Lothian. A Gamel de Linton witnessed a charter by the Bishop of St Andrews (*c.* 1160–2); Philip de Lynton was Constable of Berwick in 1285; and Thomas Lynton was Provost of Aberdeen, 1348–9. Family tradition points to a relationship with Bernard de Linton, who was appointed Abbot of Arbroath in 1311, after he had been chancellor of Scotland for some years, during almost the whole reign of Robert the Bruce. He composed a Latin poem on the victory at Bannockburn (1314), and may have fought there. He was a scholar and poet as well as churchman, warrior and statesman. He resigned the abbacy on becoming Bishop of Sodor in 1382 [1].

According to the register entry, Hercules Linton, lawful son of Alexander Linton, carpenter, and Jane Anderson his spouse, was born in the parish of Bervie (Inverbervie), Kincardineshire, on 1st January 1836 and baptised on the 15th. He was the eldest child and only son of the five children of the marriage. He was named Hercules probably after his maternal grandfather, Hercules Anderson, watch- and clockmaker in Bervie. (Family tradition has it that his mother's side would have liked him to follow his grandfather and uncle in the watchmaking business, but his father thought otherwise.)

His father was born in 1806 and may have served his apprenticeship in a small local shipyard, or more probably in Aberdeen at Messrs Hall's yard, where his son later was to be apprenticed.

He is next definitely traceable on his nineteenth birthday when by an indenture dated 1st January 1855 he was apprenticed to Messrs Alexander Hall & Sons, shipbuilders, Aberdeen [7].

According to early obituary notices, Hercules Linton rose to a position of some importance with the firm before leaving to take up an appointment as surveyor of ships. His late entry into apprenticeship may bear out the impression given in family information that he went to Messrs Hall's yard to gain practical experience before seeking to follow a career similar to his father's.

During the first quarter of the nineteenth century, the Aberdeen shipbuilding industry grew to considerable proportions. Six yards were in operation and the leading place was taken by Messrs Alexander Hall & Co., a firm which was founded in 1790 and achieved a reputation for wooden vessels. There are no records before 1811, but from 1811 to 1877, 290 ships were built, ranging in size from 20 to 2,600 tons. Their schooner *Scottish Maid* (1839), built to compete with paddle steamers between Aberdeen and London, is said to have been the pioneer of clipper-built ships, swift sailing, sharp at the bow ('Aberdeen bow'), fining away to the stern, in contrast to the bluff bows and full lines of the frigate-built predecessors.

Messrs Walter Hood & Co., established in 1839 and in existence until 1881, built some thirty-five to forty wood, composite, and iron vessels, and shared with Hall's the honour of producing clipper-built ships. Messrs Hall, Russell started iron shipbuilding in 1867 and in the following years turned out more than 100 vessels.

The fame of Aberdeen in shipbuilding was based on the great tea clippers. The early clippers of a smaller type were modelled, like the Baltimore clippers, on the French lugger. Hall's *Torrington* (1845) for Messrs Jardine, Matheson & Co., was the first British clipper to enter the China seas. In 1850 that same firm ordered a vessel with the fine lines and heavy rig of the American ships, but with superior strength in the building. The first of these Aberdeen clippers, *Stornoway* (506 tons), was followed by others, but it was not until 1852, when Hall & Co. built the *Cairngorm* (1250 tons), that a British clipper could equal if not surpass the speed of their American competitors.

CHAPTER 1

Linton is regarded as a locality name. There is a village Linton, now West Linton, in Peeblesshire; Linton a parish in Roxburghshire; and East Linton a small burgh in East Lothian. A Gamel de Linton witnessed a charter by the Bishop of St Andrews (*c.* 1160–2); Philip de Lynton was Constable of Berwick in 1285; and Thomas Lynton was Provost of Aberdeen, 1348–9. Family tradition points to a relationship with Bernard de Linton, who was appointed Abbot of Arbroath in 1311, after he had been chancellor of Scotland for some years, during almost the whole reign of Robert the Bruce. He composed a Latin poem on the victory at Bannockburn (1314), and may have fought there. He was a scholar and poet as well as churchman, warrior and statesman. He resigned the abbacy on becoming Bishop of Sodor in 1382 [1].

According to the register entry, Hercules Linton, lawful son of Alexander Linton, carpenter, and Jane Anderson his spouse, was born in the parish of Bervie (Inverbervie), Kincardineshire, on 1st January 1836 and baptised on the 15th. He was the eldest child and only son of the five children of the marriage. He was named Hercules probably after his maternal grandfather, Hercules Anderson, watch- and clockmaker in Bervie. (Family tradition has it that his mother's side would have liked him to follow his grandfather and uncle in the watchmaking business, but his father thought otherwise.)

His father was born in 1806 and may have served his apprenticeship in a small local shipyard, or more probably in Aberdeen at Messrs Hall's yard, where his son later was to be apprenticed.

In 1843, in the parish of Bervie (including the village of Gourdon) about fifty persons were engaged in agriculture, some fourteen males (masters and boys) at two spinning mills, and about 112 persons in weaving dowlas linen. There were also about forty masters—shopkeepers, shoemakers, blacksmiths, masons, occasionally employing several hands—including one boatbuilder and one watch- and clockmaker [2].

According to information in the family, Hercules was about eleven years of age when his mother died, probably in 1847, when his father was working away from home at Findhorn on the Moray Firth. At that time Findhorn, although without a harbour and with only two small stone piers, was an important trading centre, with shipbuilding and commerce with all countries, especially northern Europe and the Baltic. In 1855 there were ninety-four fishermen and a fleet of twenty-four boats; Findhorn might have developed into a herring port [3].

On 17th March 1853, Alexander Linton was appointed by Lloyd's Register of Shipping to be their surveyor for Belfast and Dublin, and held that position until 1869, when he was transferred temporarily to Glasgow. On 22nd March 1869 he had written to his head office requesting to be transferred to some port in Scotland. After a special meeting of the general committee on 6th May he was ordered to proceed to Glasgow to attend to duties there during the absence, because of a shipyard accident, of Mr Alfred Darling. At their meeting on 19th August, Linton was appointed second surveyor for the Aberdeen district (including Banff, Peterhead and Fraserburgh). After further correspondence in September (Linton's letters of 22nd and 27th September), he was ordered on 30th September to proceed immediately to Aberdeen, where he remained until his death on 6th May 1873 [4].

At about the time of his appointment Lloyd's Register had a staff of sixty-three surveyors, of whom thirteen were exclusively servants of the society (three in London, two each in Liverpool, Sunderland, and Hull, and one each for Bristol, Glasgow, Leith and Newcastle). The surveyors were of two kinds: shipwright surveyors, practical shipwrights who had served apprenticeships in the usual way; and nautical surveyors, shipmasters acquainted with the construction and repair of ships. The primary duty of the shipwright surveyors was to inspect new vessels during building. Nautical surveyors undertook the inspection of ships afloat. Both joined in the surveys of vessels in dry dock. The salaries of exclusive surveyors ranged from £150 to £500 per annum. As second surveyor in Aberdeen and district, Linton's salary was at the rate of £250 per annum.

CHAPTER 2

Information in the family suggests that the boy Hercules went to school, not in Bervie, but to the parish school in the nearby village of Arbuthnott, where the parish schoolmaster was the Rev. William Chrystall, M.A. He may have been one of a small number of boarders at the schoolhouse.

Mr Chrystall was remembered as a gentleman of high personal character and scholarly attainments, of fine physique and dignified deportment, and a painstaking teacher with a reputation for 'bringing on' lads of promise. His favourite hobby was botany, for his knowledge of which he was widely regarded. He was also session clerk and registrar, and (after the introduction of the Poor Law) inspector of the poor. He is said to have performed his duties with scrupulous fidelity and care, and the public records kept by him were noted for their neat and beautiful handwriting [5].

One of his pupils was George Brand, who began as a barefoot herd boy, worked as a mason's labourer, and with Mr Chrystall's help studied in his spare time and proceeded to the University of Aberdeen. After taking his degree he became tutor to Lord Haddo and later entered the consular service, becoming consul at Lagos.

Family information also suggests that Hercules Linton passed later to Arbroath Academy, again probably as a boarder. Unfortunately there do not appear to be any records of the Academy at about the time when Linton may have been in attendance [6].

He is next definitely traceable on his nineteenth birthday when by an indenture dated 1st January 1855 he was apprenticed to Messrs Alexander Hall & Sons, shipbuilders, Aberdeen [7].

According to early obituary notices, Hercules Linton rose to a position of some importance with the firm before leaving to take up an appointment as surveyor of ships. His late entry into apprenticeship may bear out the impression given in family information that he went to Messrs Hall's yard to gain practical experience before seeking to follow a career similar to his father's.

During the first quarter of the nineteenth century, the Aberdeen shipbuilding industry grew to considerable proportions. Six yards were in operation and the leading place was taken by Messrs Alexander Hall & Co., a firm which was founded in 1790 and achieved a reputation for wooden vessels. There are no records before 1811, but from 1811 to 1877, 290 ships were built, ranging in size from 20 to 2,600 tons. Their schooner *Scottish Maid* (1839), built to compete with paddle steamers between Aberdeen and London, is said to have been the pioneer of clipper-built ships, swift sailing, sharp at the bow ('Aberdeen bow'), fining away to the stern, in contrast to the bluff bows and full lines of the frigate-built predecessors.

Messrs Walter Hood & Co., established in 1839 and in existence until 1881, built some thirty-five to forty wood, composite, and iron vessels, and shared with Hall's the honour of producing clipper-built ships. Messrs Hall, Russell started iron shipbuilding in 1867 and in the following years turned out more than 100 vessels.

The fame of Aberdeen in shipbuilding was based on the great tea clippers. The early clippers of a smaller type were modelled, like the Baltimore clippers, on the French lugger. Hall's *Torrington* (1845) for Messrs Jardine, Matheson & Co., was the first British clipper to enter the China seas. In 1850 that same firm ordered a vessel with the fine lines and heavy rig of the American ships, but with superior strength in the building. The first of these Aberdeen clippers, *Stornoway* (506 tons), was followed by others, but it was not until 1852, when Hall & Co. built the *Cairngorm* (1250 tons), that a British clipper could equal if not surpass the speed of their American competitors.

Thereafter these Aberdeen-built ships, with their strength of build and ability to deliver their cargo in better condition, obtained the preference in the China tea trade. But clippers were not exclusively used in the tea trade. Messrs George Thompson & Co.'s Aberdeen White Star Line used their fleet of clippers built by Messrs Walter Hood & Co. to trade to Australia and all parts of the world. Once the Americans were ousted from the China tea trade the competition rested with British designers, and Clyde-built clippers—*Sir Lancelot*, *Taiping*, *Ariel* and *Serica*—were unrivalled until *Thermopylae* came from Hood's yard in 1868. Only the *Cutty Sark* (1869) disputed the *Thermopylae*'s position as the fastest sailing ship in the world. The opening of the Suez Canal in 1869 and the development of steamships scattered the clipper fleet [8].

CHAPTER 3

References to the career of Hercules Linton, after his death in 1900 (for example, the *Montrose, Arbroath and Brechin Review*, 25th May 1900) state that after leaving Messrs Hall & Co. he became an inspector in Lloyd's Liverpool office before turning to shipbuilding on his own account. He was not at any time a surveyor to Lloyd's Register.

The misunderstanding may have come about because in 1845 a Liverpool branch committee of Lloyd's Register had been set up and the reports of surveyors in the Liverpool district were dealt with by this Committee. However, from 1862 to 1885 there existed the Liverpool Underwriters' Registry for Iron Vessels. The establishment of this rival registry at Liverpool was due to various causes—dissatisfaction with Lloyd's Rules for Wooden Ships and their initial method of classifying iron ships, and possibly civic pride. The Liverpool Underwriters' Registry functioned with a fair measure of success for several years; but the Rules of Lloyd's Register were revised in 1863 and more drastically in 1870, and the amalgamation of the two registers was only a matter of time. There was much question and discussion, but the basis of agreement was not reached until 1885.

In the biennial edition of the Liverpool Underwriters' Register for 1863 and 1864, Hercules Linton is named as one of their two surveyors for the East Coast district. His name does not appear in the 1865/6 edition.

Certain early records of the Liverpool Underwriters' Association are preserved at Lloyd's Registry and from these it appears that Linton was assisting the chief surveyor (John Jordan)—as he was later described—from about January 1862. Mr Jordan, a consulting engineer, had been asked to give a full report, including a statement of fees to be received and probable income, and staff required to carry out the survey and classification of iron ships in a proper manner. His most interesting report is dated from Liverpool on 31st July 1862. The appointment of additional surveyors was recommended:

> For the East Coast it will be necessary to appoint a resident surveyor whose beat should extend from Hull to Newcastle. Mr. Linton who has been acting for me in the neighbourhood of Sunderland and Jarrow would be a most eligible young man. His experience with me for the past six months has been of a most extensive character and his abilities are of a high class. His manner is also in his favour and his strict integrity and firm way of performing disagreeable duties enables me to speak strongly in his behalf.

Linton was then serving on a monthly basis (£14 a month). Various vacancies for surveyors were advertised and Linton, then at Sunderland, aged 27, applied and was interviewed for a 'second class appointment' (£300 per annum) on 16th December 1863, but not appointed. In March 1864 he was apparently considering a change as he wrote asking what length of notice the Committee would require in the event of his wishing to leave the registry. The reply was three months, unless the committee could replace him satisfactorily in a shorter time. From a little evidence later it seems certain that he left the direct service of the Liverpool Underwriters' Association in May 1864, but continued as a 'free-lance' surveyor.

In March 1865, John Jordan, who then described himself as 'late Chief Surveyor Underwriters' Registry, now Engineer and

Shipwright National Steam Co.', wrote again on behalf of Hercules Linton, this time in support of his application for appointment as a surveyor to Lloyd's Register. Two surveyors had retired in June 1864 and after temporary arrangements, the general committee in February 1865 decided to advertise two appointments, one for the United Kingdom and one for the Colonies—candidates to be in the age range 25–35, and the salary to be £200 per annum. In April 1865 the general committee selected candidates for the two appointments, but Linton's name was not among those to be interviewed. Eight candidates for the United Kingdom appointment were interviewed in May, and in the event two appointments were made, as well as one for the Colonies.

Linton had evidently made a very serious application and had submitted eighteen statements from shipbuilding firms, shipowners, and various individuals—surveyors, inspectors, engineers—in support. A printed copy of these eighteen statements, all dated in March 1865 and referring to his application, has recently come to light. It was presented to the *Cutty Sark* Society by Miss Ann Linton in 1967.

Jordan's statement, dated from Liverpool on 23rd March 1865, is as follows:

> I am glad to be able to bear testimony to your sterling integrity, and to your skill as a shipwright in every branch of the business.
>
> As a modeller, your designs are singularly well balanced; as a judge of workmanship and material, I have met with none so much to my mind.
>
> I speak thus confidently of your capacity, after an experience of over five years, during which time you have acted as my deputy over the range of the East Coast from Hull to Kirkcaldy. In this time several hundred vessels of all sorts, viz., iron, composition (my patent), and wood vessels have passed through our hands.

Other individuals testifying were in Dublin, Jarrow, South Shields, and Hartlepool.

The following quotation from the testimonial of Messrs Shallcross and Hyam, shipowners of Liverpool, dated 23rd March 1865, is interesting because of its mention of particular vessels:

> Your abilities as a Surveyor are of the highest order, and we have every reason to be satisfied with your integrity, attention, and

skill as special Surveyor of seven of our ships recently built for us. You came to us with a character for the knowledge of both wood and iron vessels—although with us your services were specially directed to iron ships and their outfit.

The names of the vessels superintended by you are the *Hindostan, Corrientes, Carrizal, Acapulco, Surat, Iron Crown,* and *Chanaral,* and it may be interesting to Lloyd's Committee to refer to their own reports of these ships.

All seven vessels appear in Linton's own list of vessels surveyed by him, and four of them were built in the yard of T. R. Oswald of Sunderland, with whom Linton was more closely associated in later years.

Oswald's testimonial is dated 25th March 1865. The following is an extract:

It is now about 3½ years since we first became acquainted, and from that time till May last, you have continuously been surveying vessels built by my firm, and I have pleasure in stating that I have always found you master of your business, and courteous, yet firm, in carrying out the duties of your office.

I consider you well qualified to fill the post of Surveyor under Lloyds, and trust you may be successful in obtaining the appointment.

The mention of 'May last', i.e. 1864, would point to Linton's leaving then the service of the Liverpool Underwriters' Association.

The eleven shipbuilding firms that made statements on Linton's behalf were in Belfast, Dublin, Newcastle, Middlesbrough, Stockton-on-Tees, Sunderland, and Hartlepool. Messrs Harland & Wolff of Belfast testified as follows:

We have known Mr. Linton for many years, during a portion of which time he was in our employment—first working practically in the yard and afterwards in the model-room; and we found him during that period, and on every other occasion of meeting him, straightforward, intelligent, and industrious, and feel sure that if he obtained the situation for which he is applying, he will give every satisfaction to his employers.

Linton's father, Alexander, was a surveyor to Lloyd's Register who had been stationed in Belfast since 1853. It is not yet known when Hercules Linton was employed by Harland &

Wolff, although the family must have been known to the firm. He might have been employed before 1st January 1855 when he entered on his apprenticeship with Messrs Alexander Hall & Sons, Aberdeen, or more probably after he had completed his apprenticeship in December 1860. This might square also with Jordan's statement in 1865 that he had known Linton for more than five years. Jordan no doubt had contact with Harland & Wolff, and with Linton, before Linton began assisting him in the survey of vessels early in 1862.

It is a little puzzling why Linton was not interviewed by Lloyd's Committee in 1865. There is a suggestion that as he served his apprenticeship in wooden shipbuilding he might not have had sufficient basic experience in iron shipbuilding. His period in Harland & Wolff's iron shipbuilding yard may have been intended to remedy this shortcoming. His father's position may have been a complication, and there may have been other factors not now known. Linton was then twenty-nine years old.

On 12th April 1865 he married at Kinghornie (near Bervie) Marjory Anderson, 'eldest daughter of the late David Anderson, farmer'. He is described in the marriage certificate as 'surveyor of vessels', Newcastle, and in the newspaper announcement as of North Shields. Their first child, a daughter, was born in Newcastle in June 1866 and the father was still described as 'surveyor of ships'.

In an old notebook, still preserved, are listed the names of fifty-five vessels 'surveyed by H.L.'. This notebook, in Linton's own handwriting, seems to contain information entered up in fair copy. The vessels are of different kinds—sailing ships, screw steamers, paddle steamers, etc.—many of them built about the time when he was engaged on survey work. Occasionally later notes in ink or pencil have been added, e.g. '*Acapulco* Barque, lost 1876', '*Royal Victoria* do. [i.e. ship] Lost'. In a different part of the notebook are numerous careful sketch plans of constructional details in the ships surveyed, e.g. s.s. *England*—shank stopper, bowsprit cap, lift block, fife rail and goose, screw fidd, pantry (this last dated November 1865), tressel trees, topsail yard and truss, and sizes of running chain. In all there are twenty-nine sketches relating to thirteen ships.

If Hercules Linton continued as a free-lance surveyor of ships, he most probably combined his survey work with the designing of ships, in the Glasgow area. His second child, also a daughter, was born in Maryhill (Lanark) in February 1868 and his next children, sons, were born one in Dumbarton in October 1869, one in Glasgow (Govan) in January 1872—he died only nine months later—and the third in Glasgow (Tradeston) in August 1873.

CHAPTER 4

Among family papers is a business card of Hercules Linton, ship- and yacht-builder—models, designs, specifications for ships and steamers—joiner, spar maker, blacksmith, carpenter, with an office at 17 Oswald Street, Glasgow. This may refer to his first period in Glasgow, before he entered into partnership with Scott and before the firm leased the Dumbarton shipyard in May 1868. After the collapse of the firm in the autumn of 1869, he moved to Aberdeen for some months and returned to Dumbarton and Glasgow for a year or two before transferring to Dundee. He returned, however, to the same area of Glasgow sometime after 1877 until 1880.

It is very probable that Linton in his spare time as a surveyor of ships was undertaking certain private work of planning and design. Again among family papers is a book of newspaper cuttings, unfortunately without any indication of the periodical or date. Most cuttings refer to antiquarian matters—one of Linton's main interests later—but some few to family matters and ships. Two cuttings relate to a new steamship for the port of Dundrum (N. Ireland), to sail between Dundrum and Whitehaven, the *Lady Alice Kenlis*, for Messrs Hugh Andrews & Co. of Belfast—a sister ship for the *Lady Alice Hill*. The *Lady Alice Hill*, an iron screw steamer (258 tons), was built by Denton of Hartlepool and launched in August 1866. The *Lady Alice Kenlis*, one of the ships surveyed by Hercules Linton, was built by Swan of Maryhill (Glasgow) 'from designs by, and under the direct superintendence of Mr. H. Linton, C.E., son of Mr. Linton, Lloyd's Surveyor for the port of Belfast, and which reflect great credit on that young gentleman's skill and ability'.

The vessel was launched on 23rd December 1867 after having been about six months on the stocks; she made ten knots on her trial trip.

The ship's dimensions were given as follows: extreme length, 130 feet; beam, 19 feet 9 inches; depth, 9 feet 10 inches; gross tonnage, 213; registered tonnage, 145. The vessel was described as of very handsome appearance and well fitted for her proposed trade—traffic in cattle, goods, and passengers. Her build was said to be of extraordinary strength, 'similar to the gunboats of her Majesty's navy'. The entire shell was one-sixteenth of an inch thicker than required by Lloyd's Rules:

> twenty tons more iron being used in the construction than specified by Lloyd's. It is fitted with a strong box keelson, twelve inches by ten, two bilge angle irons and side girders, with double reverse bars running to the upper turn of the bilge, and reverse bars running alternately to the gunwale . . . the keel, stem and stern posts . . . larger than required by Lloyd's rules. An important feature in the construction is the existence of two outside keels running the entire length of the flat of the floor, which will enable the vessel to take the ground with greater safety. The engines were supplied by the Greenock Foundry Company, and are built on the direct-acting condensing principle. They are nominally of 40 horse power, and are fed by two tubular boilers with two furnaces. The engine-house and boilers are secured by a strong teak-wood frame, water-tight, with skylight, which will be found very serviceable in heavy seas by protecting the fires. There is also a donkey-engine for supplying steam to the windlass, situated abaft the foremast, for the discharge of cargo and the loading of cattle and merchandise—the smoke being conveyed by pipes into the main funnel, so as not to obstruct the view of the Steersman. The ship is rigged as a fore-and-aft three masted schooner, with all the modern appliances, including a steam patent windlass for lifting anchor. The three masts are of Savannah pitch pine, of one length from keel to trunk, and the compasses are of the finest description—similar to those used in the West Indian and Pacific mail steamers.

The newspaper reporters—Scottish and Irish—were evidently well briefed when they were able to look over the new screw steamer after she had arrived in Belfast Lough on Saturday morning, 14th March 1868, and had been moored in the Abercorn basin. The *Lady Alice Kenlis* had made the voyage from Ardrossan the previous night, with 260 tons of coal, in eleven hours forty minutes in the teeth of a strong head-wind.

In the afternoon of that Saturday, 14th March 1868, a number of gentlemen were invited to lunch on board the vessel and a toast was proposed for the success of the *Lady Alice Kenlis* and her worthy owners, Messrs Hugh Andrews & Co. Mr Andrews, replying, referred to 'their talented young friend, Mr. Linton, who seemed to have inherited the talent of his father, who was so well known to them'.

Shortly before Hercules Linton (then aged thirty-two) joined William Dundas Scott (aged twenty-three) in partnership as shipbuilders, letters passed between them in March 1868, and drafts or copies of two of these, from Scott to Linton dated 14th and 17th March, are extant [9]. In his letter of 14th March, Scott, among other matters, hopes that Linton will have a pleasant passage with the little ship to Belfast. In his letter of 17th March, Scott states:

> I am disappointed at your being delayed so long in Belfast. I intend going home tomorrow to Roxburghshire, where I will expect you to join me on your return from Ireland. We can consult with my Father at home about the amount of capital required, etc., and then return to Glasgow where we will have a deed of partnership drawn out. We will then buy our plant and launch ourselves forth as a firm of Clyde Ship Builders.

The newspaper cutting referring to the *Lady Alice Kenlis* and the voyage from Ardrossan to Belfast on a Friday night, is without date and without reference to a date, but the two letters from Scott to Linton leave no doubt that the Friday night was the night of Friday, 13th March 1868, and the account of proceedings in the Abercorn Basin, Belfast Lough, given above, is therefore dated Saturday, 14th March.

Scott's letter of Tuesday, 17th March 1868, contains the first reference to the plans of Scott and Linton to set up in partnership as shipbuilders, and from later information about him, it might be supposed that although Scott was the younger man, his enthusiasm may have led him to be somewhat over-confident, over-optimistic, and over-persuasive. The deed of partnership or articles of association have not survived. The firm was not registered with the Registrar of Companies in Edinburgh, or mentioned in the records of the Glasgow Chamber of Commerce. However, in connection with their later application for discharge from bankruptcy, both partners were examined before Mr Sheriff Steele in the Dumbarton bankruptcy court on Monday, 16th January 1871, and accounts of the proceedings were given in the two local weekly newspapers: the *Dumbarton Herald*, Thursday, 19th January 1871 (p. 1), and the *Lennox Herald*, Saturday, 21st January 1871 (p. 4) [10]. Scott's sworn testimony was as follows (Hercules Linton's statement was said to be to the same effect):

At Whitsunday, 1868 [i.e. 31st May], I entered into business as a shipbuilder at Dumbarton, in partnership with Hercules Linton, the firm being Scott and Linton. Our capital at that time was £600, afterwards increased to £1700 [this seems to be a mistake for £1,200]. Of this I received £600 from my father, and of the remainder, £300 stood on a cash credit in the Royal Bank, and the other £300 was borrowed on a policy of life insurance. [Linton had a policy for £500 —premium £6. 17*s*. 1*d*. per annum. See McCall letters—John Wild to Scott, 14th September 1869.] We had no liabilities at the time we commenced business. Mr. Linton managed the shipbuilding and I attended to the counting house and the engineering. The company were getting into a state of embarrassment shortly before September 1869, and we accordingly erected [*sic*] a trust deed in favour of Mr. Matthew Paul and others, for behoof of our creditors. Our embarrassment arose from our inability to execute the orders undertaken in time, in consequence of want of machinery. The instalments which ought to have been due were thus retarded, and we were consequently unable to meet our current engagements. We had made no balance of our books on our commencement, till immediately preceding the date of the trust deed, and when we did then balance our books the result of that investigation was to show that the business had not been profitable, and that the winding up of the business would show a probable loss of £7,000. A trust deed was then granted, and thereafter sequestration was applied for. We granted no preference to any of our creditors, and we surrendered to our trustee our whole assets for behoof of the creditors.

The McCall documents (see note 9) give certain revealing and personal insights into the relationships of the two partners, especially at the time of the bankruptcy. There are thirteen letters from Linton to Scott (twelve of them in the two years

May 1868 to May 1870), drafts of two letters from Scott to Linton in March 1868, drafts or copies of various business letters from Scott (including one to his father about a loan), various other business letters to Scott, together with receipts, bank statements and so on. It would seem that Scott was the business secretary while Linton was the commercial traveller, using his earlier contacts as surveyor of ships to get orders (mainly in Liverpool and London), but both lacking business experience. In March and April 1868 there were inquiries about a suitable yard, at Port Glasgow and even near Port Edgar (the North British Railway terminal on the Firth of Forth) before they settled to rent part of Denny's yard (the Woodyard) at Dumbarton at £200 per annum from 15th May 1868. In that same month Scott was writing for advice on money matters, explaining to James Mylne that his father was highly pleased with the yard, but all at sea on the subject of raising money, and already there was a small wages bill for a few men preparing the first keel and a few cash payments to be met without delay. In August 1868, according to Scott's draft of a letter to his father, they were much in need of £75 and an advance of £200 to complete work on a steamship, *Camel*, for Bibby's, ordered to be completed in seven weeks' time, at a price of £980, cash on delivery at Liverpool. Scott explained that they did not wish to ask the bank to advance the money and as their estimated profit was £250 he suggested a loan of £200 for two months in addition to the £75. (The Commercial Bank of Scotland's statement to 31st December 1868, for Scott & Linton, shipbuilders, Dumbarton, showed a credit balance of £24. 19*s*. 5*d*.)

An example of difficult working between the partners may be cited. Linton wrote from Liverpool on 29th May and again on 9th July 1868 about the possibility of an order for an 800-ton sailing ship for a Mr Preston at £14 a ton—half cash on launching (preferably by instalments as the work proceeded) and half by bill at six months—and asked for a reply by telegram. In his letter of 13th July (from London) Linton said that the telegram came too late to be of use and that Dobie's, a competing shipbuilding firm, took the order. Again, in the draft or copy of Scott's letter dated 4th February 1869 to his namesake Walter Scott of Tufnell Park Road, London, N.—apparently about abortive negotiations, 'Mr. Willis' ship [*Cutty Sark*] having come in the way of yours'—there is the revealing sentence, 'I must really plead inexperience more than anything else as an excuse for having been so long in the belief that we would not be losers by your proposed contract.' It seems clear that the partners were anticipating and counting on the instalment payments on the *Cutty Sark* when discussing or negotiating the building of the other vessel.

In the old notebook previously mentioned is a list of nine ships 'Built & Designed by H. Linton':

Camel, steamship, owners J. Bibby & Co. Price £980.
Jura, steamship, owner A. McKechnie. Price £1,275.
Tug, twin-screw, owner A. Leslie. Price £620.
Exploring composite twin-screw, sent to Japan in pieces.
Snowdon, paddle steamer, owners W. R. Preston & Co. Price £2,700 not including old steamer taken in part payment.
Giao, paddle steamer, owner Donna Maria. Price £2,225.
Invereshie, iron sailing ship, owners I. & K. Grant. Price £10,700.
Cutty Sark, composite sailing ship, owners J. Willis. Price £16,150.
Linfern [*Linn Fern*], composite sailing three-masted schooner. Price £2,475.

The completion of the steamship *Camel* in October 1868 has already been mentioned, and it is reasonably certain that all the other vessels came from the Dumbarton yard. The activities of the young firm can be followed in some detail, not only from references in the McCall papers, but especially in the columns of the two local weekly newspapers. The *Dumbarton Herald* appeared on Thursdays, and the *Lennox Herald* on Saturdays, and both papers were published by the same Dumbarton firm of Bennett & Thomson. Both papers made mention from time to time of shipbuilding matters in Dumbarton, as well as elsewhere, and there was much duplication of local information. At the end of December or the beginning of January in each year both papers gave a review of shipbuilding at the different yards in Dumbarton—launches during the year, work in progress, and contracts and prospects for the coming year. Moreover, four of the vessels—*Snowdon*, *Invereshie*, *Cutty Sark*, and *Linn Fern*—were classed with Lloyd's Register, and there is thus more information about them [11].

On the general side, it was reported in January 1869 that

Scott & Linton 'since they entered, some six months ago, on the occupancy of the lower portion of the yard on the west side of the Leven', had done little more than get things into working trim. Only one vessel, the *Camel* (a small iron screw steamer) had been launched in 1868, but a list of six ships—an iron sailing ship, two composite sailing ships, an iron paddle steamer, a composite twin-screw steamer, and an iron screw steamer—was given as being built or contracted for in 1869.

In April 1869 the whole of the shipbuilding yard—the Woodyard—was leased 'since their orders in hand made this necessary'. Yet a meeting of the principal creditors of Scott & Linton was called for Friday, 10th September. Work in the yard had been partially suspended for several days before. At the meeting of creditors, trustees were appointed to make arrangements for the completion of the firm's contracts.

In January 1870, the *Lennox Herald* expressed regret that the firm of Scott & Linton had been obliged to abandon operations in the Woodyard and gave a list of the vessels launched in 1869—*Snowdon*, *Jura*, *Giao*, *Invereshie*, a screw launch, and the *Cutty Sark*. Only one vessel was mentioned as being built or contracted for—a composite sailing schooner.

The financial difficulties over completing the *Camel* have already been mentioned. This small iron screw steamer, built to the order of Messrs John Bibby, Sons & Co., of Liverpool, and intended for the Mediterranean trade, was launched on Saturday evening, 10th October 1868, in the presence of a number of the friends of the builders, and ladies and gentlemen of the locality. She was named by Miss Helen Gibson, daughter of Archibald Gibson, Esq., secretary to the Caledonian Railway Company. After the launch, the invited guests partook of a 'sumptuous and recherché' meal (provided by Messrs Ferguson and Forrester of Glasgow) in the 'draughting room'. Scott was in the chair, and the health and prosperity of the firm was proposed by Mr Gibson. Scott in his reply expressed the firm's wish for good relations with their employees. They had begun their work under the shade of a name (Denny's, who had the Woodyard before them) made great in shipbuilding, and hoped to merit the esteem of all and the patronage of the shipbuilding world. Other toasts and replies followed. Perhaps this entertainment was thought to be good publicity.

The mention of the launching of 'Leslie's little boat tonight', in Linton's letter of 7th April 1869, may refer to A. Leslie's twin-screw tug (price £620).

The paddle steamer *Snowdon* (ship no. 2 according to the yard number from Scott & Linton's yard) was under special survey for Lloyd's Register during the whole time she was being built, from 15th January to 12th May 1869. The surveyor, John F. Light, later assigned to the Liverpool district, paid twelve visits while the steamer was on the building slip and afloat. The *Snowdon* was launched on Friday, 16th April 1869, and named by Miss Linton, one of Hercules Linton's sisters. The ship, for the Liverpool, Rhyl & Vale of Clwyd Steam Packet Co. for service between Liverpool and Rhyl, and to carry passengers as well as freight, was to have a handsome saloon and forecabin. Her gross tonnage was given as 160.4, registered tonnage 94.12; her two engines developed 80 h.p. She was expected to have great speed and to be superior to those vessels hitherto employed in the same trade.

On Monday, 24th May 1869, a 'smart' screw steamer of 120 tons burthen was launched from Scott & Linton's yard. Her measurements were: length 86 feet; breadth $17\frac{1}{2}$ feet; depth $8\frac{1}{2}$ feet. She was intended for the West Highland trade and named *Jura* by Miss Mackechnie, of Shields Terrace, Glasgow, the daughter of the owner.

I have not yet discovered when the twin-screw composite exploring ship was built, for sending to Japan in sections, but on Wednesday, 9th June 1869, a substantially built smart paddle steamer of about 100 tons B.M. was launched from the yard. She was schooner rigged, without bowsprit, 100 feet in length, $14\frac{1}{2}$ feet beam, and 6 feet depth. She was built to the order of a Brazilian firm (per Mr Tappenbeck of Liverpool) and intended to be the first of a fleet of similar vessels to ply on the river Amazon. Messrs W. M. Robertson & Co. of Glasgow were to supply the diagonal engines. The vessel was named *Giao* by Mrs. Rich, the wife of Captain Rich who was to take command of her.

'Immediately after the launch, the place which the *Giao* had occupied was cleared, and the blocks laid for the keel of an iron three masted schooner of 300 tons, for Glasgow owners, and intended for the West India trade.' (This was probably the three-masted composite schooner, the *Linn Fern* of about 350 tons.)

The iron sailing ship *Invereshie* (ship no. 7) was built under special survey between 18th June and 27th December 1869, for Messrs J. & R. Grant of London, 'to class Aᴀ at Lloyd's and 20 years in the Liverpool books'. Her length was 185 feet, depth 32 feet, gross tonnage 742.39, under-deck tonnage 687.01. The under-deck tonnage was less than anticipated and the scantlings therefore in excess. As the vessel was to be one of an East India fleet and to use the Suez Canal, provision was made for auxiliary steam power. The stern post had been formed for the purpose of receiving an auxiliary screw propeller, but plans had been changed and the aperture filled up and plated over. The vessel was launched on Saturday, 6th November 1869, and named by Miss Evelyn Grant.

The First Entry Report on the *Invereshie* for Lloyd's Register, received in London on 13th January 1870, was signed by two surveyors, Alexander Linton and W. T. Mumford. It will be remembered that in May 1869, Alexander Linton, serving in Belfast (with Dublin), had been asked to proceed to Glasgow to take over from Mr Alfred Darling who was on sick leave. Linton transferred to Aberdeen at the end of September 1869, but in August he was joined by Mr W. T. Mumford from Liverpool and a new appointment, Mr William Moverley. The general committee of Lloyd's Register, at their special meeting on 19th August 1869, in reply to a memorial from the shipowners and shipbuilders in Glasgow, agreed that Mr Darling, who had suffered an accident in McMillan's shipyard, Dumbarton, and was then in London, should return to Glasgow.

The keel of the *Cutty Sark* may have been laid in February 1869, shortly after the agreement had been signed on 1st February. Completion was due on 30th July 1869, but special circumstances in connection with the building of the 'clipper' ship caused delays and the vessel was not launched until 22nd November and not completed until January 1870. Because of these special circumstances it will be convenient to deal with the last vessel to be launched from Scott & Linton's yard before returning to the complicated history of the *Cutty Sark*.

A three-masted composite schooner of about 350 tons was launched on 18th January 1870 and was thus on the stocks for some months at the same time as the *Cutty Sark*. The vessel, described as a fine model and expected to be a fast sailer, was built for Messrs J. and R. Neill, shipstore merchants, and Messrs D. Whyte & Co., shipbrokers, Glasgow, and intended for the South American trade. She was named the *Linn Fern* by Miss Isabella Neill, the daughter of ex-Baillie Neill of Glasgow.

Linton's letter of 20th January 1870 to Scott, from Aberdeen after a recent visit to Dumbarton, refers not only to the *Cutty Sark*'s sailing to London, but also to the schooner (*Linn Fern*) as a 'perfect beauty', although the builders had a great deal of trouble because her captain objected to almost everything that was done. In his letter of 26th January 1870 from Aberdeen, Linton mentioned that the *Invereshie* 'arrived in London last week', having made the passage in four days, a fine run.

The *Cutty Sark* was to be built under special survey and classed A.1. at Lloyd's for sixteen years. The First Entry Report was signed by the same two surveyors, A. Linton and W. T. Mumford, who signed the report on the *Invereshie*. No date was given for the first survey visit, but the last took place on 7th January 1870; the ship was launched on Monday the 22nd November 1869. In their general remarks, the two surveyors refer to the fitting of additional bilge plates (10″×10/16″, from bulkhead to bulkhead) and five beams at the fore-end and four at the aft, to prevent 'panting', as sanctioned by Lloyd's Committee in their letter of 28th August 1869. These remarks call for explanation and amplification since they are closely bound up not only with the building of the *Cutty Sark* but with the fortunes of the shipbuilding firm of Scott & Linton in its final liquidation; the *Cutty Sark* was their fifth vessel, and their first to approach 1,000 tons in size.

Basil Lubbock gave a very full and at times circumstantially convincing account of the building of the ship, but this now needs correcting and supplementing by information which has recently come to light [12]. A summary only is attempted here.

In the late 1850s and in the 1860s it was the ambition of shipowners to win the Great Ship Race from China with the first teas of the season. Captain John (Jock) Willis had been unsuccessful, but was keen and persevering. He had commanded his father's ships when American clippers carried all before them, and in the early 1860s his *Whiteadder* was outclassed by the Clyde-built clippers. Then in the autumn of 1868 the *Thermopylae* sailed from London on a record-breaking trip; but Willis was determined to beat her, and had discovered Linton of Scott & Linton of Dumbarton. Willis was also the owner of the *Tweed*, the flagship of his fleet, which he considered to be the fastest sailing ship ever built, although too big for the tea trade. The story is that Willis took Linton to see the *Tweed* in dry-dock; her master Captain Stuart and former mate, Captain Moodie, who was to command the new clipper, were also present, and the lines of the *Tweed* were to be a guide for the new ship. The late Mr J. M. Murray questioned how much the *Cutty Sark* was influenced by the *Thermopylae*, which was built one year earlier in Walter Hood's yard in Aberdeen, from designs by Bernard Waymouth, who at that time was senior ship surveyor in London to Lloyd's Register; he apparently designed ships in his spare time. Mr Frank Carr (*Royal Institution of Naval Architects Quarterly Transactions*, July 1966, pp. 214, 215) has drawn attention to a valuable comparison between the lines, plans and sail plans of the *Thermopylae* and the *Cutty Sark* in an article by Mr Leslie Knopp in *Yachting Monthly* (February 1930). In 1867 appeared Lloyd's Rules for Composite Ships—largely the work of Bernard Waymouth—accompanied by drawings from his colleague Harry John Cornish. So far as I am aware, no one has yet compared Waymouth's Rules and Cornish's drawings with the plans, drawings and specifications of Linton for the *Cutty Sark*. Incidentally, Cornish was one of two Lloyd's Register surveyors in Glasgow in the early part of 1869.

According to Captain Moodie's working copy of the agreement between John Willis & Son and Scott & Linton for the building of the composite sailing ship, the vessel was not to exceed 950 tons; length of keel and forerake 210 feet; beam 36 feet extreme; depth 20 feet 9 inches. The model was to be prepared by Scott & Linton in accordance with the drawings and plans, and approved by Willis or his appointed surveyor. The vessel was to be built under the immediate superintendence of this surveyor (who was Captain George Moodie), who was entitled to examine the vessel, materials and workmanship, and progress of the work, but not so as to impair or diminish the responsibility of Scott & Linton for the proper execution of the work. Detailed provisions were included about alterations and additions and the insurance of the vessel until final delivery. The completed vessel was to be delivered at Greenock or Dumbarton on or before 30th July 1869. If Scott & Linton failed to deliver on time, they were to pay £5 a day compensation for each day of delay after 30th July, unless the delay was caused by agreed alterations or by strikes in shipbuilding yards.

Payment was to be made by Willis at the rate of £17 a ton—but there would be no extra payment if the tonnage exceeded 950—that is, a total payment of £16,150, in eight instalments:

£500 on the execution of the agreement.

£1,500 when the keel was laid and the stem and stern posts properly in place.

£2,000 when the vessel was in frame and the beams riveted.

£1,000 when the vessel was ready for planking.

£2,500 when the vessel was completely planked and decked.

£2,500 when the vessel was caulked, coppered, and ready for launching.

£2,500 when the vessel was launched.

The balance was to be paid on completion to the satisfaction of Willis or his surveyor—'with sails bent and compasses adjusted'.

If Scott & Linton were unable to complete, Willis had power to enter the shipbuilding premises at Dumbarton and cause the work to be completed, using materials and machinery as necessary, and making payment out of the unpaid instalments of the contract price.

The agreement was drawn by Messrs Burns & Alison, writers (i.e. solicitors), Glasgow, signed by Scott & Linton on 29th January 1869, and by John Willis & Son in London on 1st February following.

Lubbock (*The Log of the Cutty Sark*, 2nd ed., p. 24) gives an imaginative reconstruction of the first step in the designing of the *Cutty Sark*, which is worth quoting:

When the model was finished, the designers and Lloyd's surveyors, Cornish and Bernard Waymouth, probably the foremost men in their profession, went to see it and they were both aghast at the knife-like entrance. They thought there was too much cutaway, that she required more bearing forward and would therefore lack power. But they were both wrong and Linton was right. The square bilge and powerful quarters counteracted the lean bow, and the *Cutty Sark* proved herself to be a magnificent sea boat.

John Willis, who is said to have had a keen eye for a good model, approved this as well as the drawings and specifications [13].

According to Lubbock, shrewd old Willis required a guarantee that the work on the new clipper would be properly done to the requirements of the specification, and it is said that Mr Peter Denny of Messrs William Denny & Bros offered to become surety at the price on the contract, and so the matter was settled. No evidence has come to light of any such guarantee, but there were trade and financial arrangements between William Denny & Bros and Scott & Linton, as will be explained later. It is highly unlikely that the firm of William Denny & Bros did more than complete the *Cutty Sark* and other vessels at the request of, and by arrangement with, the creditors.

Willis appointed Captain George Moodie, who was to command the new clipper, to superintend its construction, and he was said to have been so particular about the materials used that, for instance, any timber with knots, shakes, or sappy patches was promptly rejected.

Captain Moodie's vigilance may have been one of the causes of Scott & Linton's financial difficulties and later bankruptcy. They were a small, young firm, without financial strength, and they encountered other difficulties before the vessel could be launched.

Filed with the First Entry Report on the *Cutty Sark* are nine letters or copies and drafts of letters which passed between the London office of Lloyd's Register and their Glasgow surveyors in July and August 1869. The correspondence is incomplete, but the sequence of events can be traced. According to the draft of a letter to Alexander Linton dated 21st July 1869, there was complaint from Lloyd's about the considerable delay in Scott & Linton's furnishing information about the *Cutty Sark,* and although the local surveyor (Mr A. Darling, later absent on sick leave) had given them early warning that the position of the bilge plate was much too high, the builders had proceeded with the planking of the bottom of the vessel. Lloyd's chief

Hercules Linton's signed and dated pencil drawing of the Woodyard and Dumbarton Rock, showing the *Cutty Sark* on the stocks (left centre).

The *Cutty Sark* in full sail, by F. Tudgay. Signed and dated 1872.

Previous page: Alexander Hall's yard, Aberdeen, 1862 (see note 7).

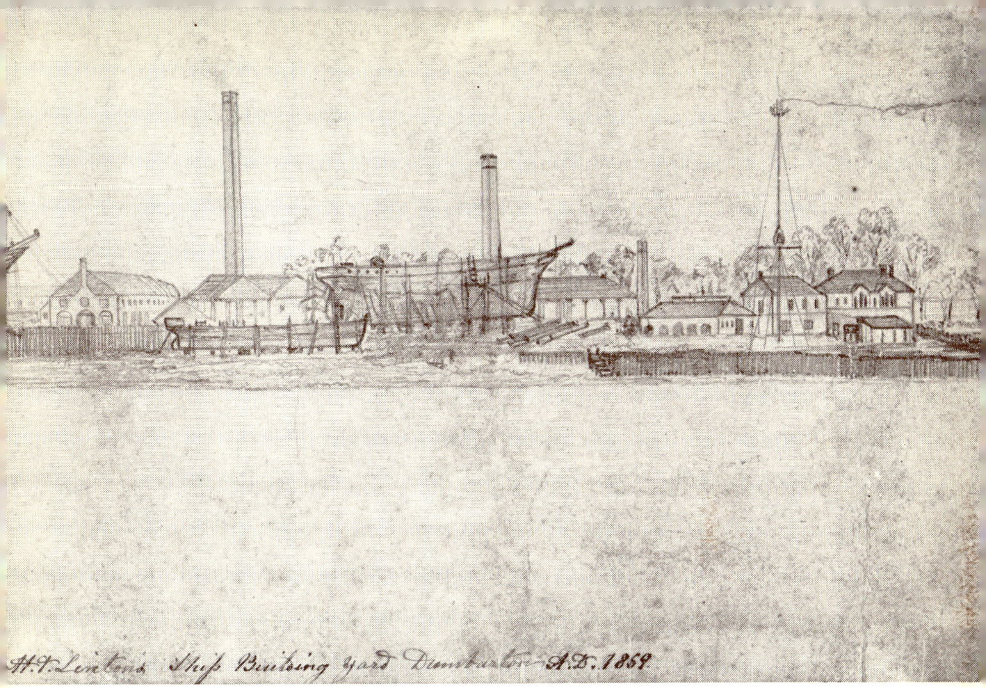

The view is from the east bank of the river Leven, looking south-west. The river turns west below the Rock and then south to join the Clyde.

Original drawing by Hercules Linton for the *Cutty Sark's* figurehead.

Following page: The *Cutty Sark* at anchor in Cork Harbour (Cobh), 1871.

surveyors—Joseph Horatio Ritchie and James Martin, both assigned to London—were of the opinion that the only way to secure longitudinal strength was by removing sufficient of the bottom planking to allow additional iron bilge plates to be fitted. Mr Darling (then in London after his accident) concurred in that opinion, and as he was to visit the Clyde he could arrange for this modification.

On 23rd July, Hercules Linton, writing from Dumbarton on behalf of Scott & Linton and referring to the letter from Lloyd's of 22nd July about additional bilge plates, pointed out that the plans of the vessel had been submitted to Mr Light (who earlier was Lloyd's surveyor in Glasgow and had signed the First Entry Report on the *Snowdon* in May 1869) who had not called attention to the requirement, or otherwise an alteration would have been made. At the present stage of the work, with the planking of the bottom all fitted and bolted, the adoption of the suggestion was precluded, although the builders would consider any other practicable suggestion.

On 24th July 1869 the two surveyors in Glasgow (A. Darling and Alexander Linton), after a survey of the vessel, 'respectfully' suggested that if the bulb bar ($10'' \times 10/16''$) were extended fore and aft, this would provide additional strength, although not equal to the additional bilge plates, but sufficient 'compensation' in the circumstances.

The opinion of the London surveyors was sought on 26th July, given on 30th July, and conveyed to the Glasgow surveyors on 31st July.

The London surveyors had also been in touch with Mr Light, who made it clear that he had not sanctioned the improper position of the bilge plate,

> and as Mr. Darling objected to its position in good time, it remains that the builders are themselves to blame for this deviation, and the consequent shortening of the diagonal straps on the frames, which are so necessary in Composite Ships to prevent torsion—and in our opinion and the opinion of *the Owner*, the extension of the bulb bar keelson to the ends of the vessel will not be of equal service to her as compensation.

They repeated their opinion that an additional bilge plate ($24'' \times \frac{5}{8}''$) should be fitted all fore and aft, well riveted to the frames, and placed where marked in red ink on the sections (now not available). Hercules Linton, on behalf of the builders, still demurred (letter dated 25th August) to compliance with the requirements of Lloyd's Register Committee when these were made known to him and confirmed again in their letter of 28th August. And yet according to the surveyors' remarks on the First Entry Report, a compromise was agreed on 28th August that extra bilge plates of smaller size (10″×10/16″ instead of 24″×10/16″) should be fitted from bulkhead to bulkhead, as well as five beams in the fore end and four aft to prevent 'panting'.

It was a pity that Alexander Linton, the father of Hercules Linton, should have been concerned with this disagreement, especially as, while serving as Lloyd's surveyor in Belfast, he was asked in May 1869 to undertake duties temporarily in Glasgow during the absence through illness of Mr Darling, and left Glasgow at the end of September 1869 to take up his new post in Aberdeen. There can be no doubt however that, because of the disagreement with Lloyd's, Scott & Linton suffered serious delay, with the attendant expense of additional work and material, postponed instalment payments, and demurrage costs. Their difficulties were manifest before the end of August and work on the *Cutty Sark* and in the yard generally was partially suspended in the first week of September. At a meeting on Friday, the 10th September, a committee of Scott & Linton's creditors decided to complete the contracts in hand. Not all the creditors were apparently informed of the meeting, and Messrs Alexander Mathieson & Sons (Saracen Tool Works, Glasgow) on 24th September wrote complaining about this. Among the McCall documents is a brief letter to Scott from William Ross, Leven shipyard, 6th September 1869, stating that after a second perusal of Scott's rough statement of their affairs, and 'taking the most favourable view, you cannot be less than £5000 short'. The statement which was attached does not now seem to be available although there is an undated statement showing a different state of affairs, with instalment payments to be expected on three ships (nos. 7, 5 & 9 i.e. *Invereshie*, *Cutty Sark*, and *Linn Fern*) and with outgoings to be met between August and December 1869. Scott's testimony (previously quoted) in January 1871, in connection with proceedings for their discharge from bankruptcy, put the probable deficit at £7,000.

The memorandum of agreement between Willis and Scott & Linton stated that if the builders were unable to complete the vessel, Willis or his nominee or nominees could enter upon and use the shipbuilding premises at Dumbarton and cause the work to be completed by persons whom he saw fit to employ. It would appear that Willis was not called on to make other arrangements since the creditors decided to complete the contracts in hand themselves. In the First Entry Reports of the *Invereshie*, the *Cutty Sark* and the *Linn Fern*, Matthew Paul and John Henderson described themselves as acting for Scott & Linton's creditors and Matthew Paul as 'Chairman of the Committee of Scott & Linton's Creditors'. They arranged with the firm of William Denny & Bros to complete the ships, and Peter Denny's assistant manager, William Ross, at the Leven shipyard, Dumbarton, acted for the creditors in 'finishing (or rather managing the finishing) of the vessels on the stocks'. The *Cutty Sark* was launched by Mrs Moodie (wife of Captain Moodie) on Monday, 22nd November 1869, and the *Linn Fern* on Tuesday, 18th January 1870. The *Cutty Sark* was later moved to the Leven shipyard for masting and then on 20th December towed to Greenock for the running rigging to be put in by a specialist firm; she left Greenock for London on 13th January 1870 and a fortnight later began taking in her first cargo at the East India Docks.

Lubbock commented that the course taken by the creditors in finishing the *Cutty Sark* at their own expense ended disastrously for them. After realising all assets, another four shillings in the pound was added to their claims. It is said that one creditor, who was owed £4,000, found that finishing the two vessels cost him another £800. The main creditors are not known, but the firm were concerned with more than two vessels. Matthew Paul, Provost of Dumbarton at the time, was the builder of marine engines who tendered for and presumably supplied engines for screw and paddle steamers. Linton in a letter to Scott, dated 26th January 1870 from Aberdeen after a visit to Glasgow, was of the opinion that their creditors were then aware that they had made a mistake in not allowing Scott & Linton to complete the ships. Paul was described as a tool in Denny's hand and Denny the main architect of their downfall. 'It was Denny's intention to make the estate pay as little as possible . . . and Denny an elder of a Church . . . who contributes to the building of churches and at the same time calculates on the downfall of his fellow men.' This bitterness towards Denny is repeated in Linton's letter to Scott of 13th May 1870. The circumstances which gave rise to it are not known; but if Denny had given a guarantee to Willis about the completion of the *Cutty Sark* and had refused to exercise patience or extend credit, he might have been held by Linton to have been a main cause of their bankruptcy.

The business records of William Denny & Bros Ltd, Dumbarton, are now preserved in the Adam Smith Building (Department of Economic History) of the University of Glasgow. An examination of the journals and ledgers of the appropriate dates shows business transactions such as might be expected between a smaller and a larger shipbuilding firm, and a cash loan of £1,000 to Scott & Linton on 20th August 1869, repaid presumably by Scott & Linton's creditors, although the firm is always referred to as Scott & Linton, on 4th October 1869. For example, Scott & Linton purchased from William Denny & Bros a punching machine (£71) in May 1869, and another one (£79. 8*s*. 10*d*.) in June, iron for bolts in July, and timber (£405. 6*s*. 10*d*.) in November. By 30th April 1870, the accounts were straight except for an amount of £159. 6*s*. 3*d*.—'amount at debit carried off, being for machinery sold them and unpaid that firm having become bankrupt' [14].

The accounts of Scott & Linton's creditors are not now available. Some indication is given in Linton's letter to Scott from Aberdeen, 20th January 1870, of the losses involved in their failure to meet completion dates for ships, with consequent retarded instalment payments and demurrage charges. Linton had heard that the creditors had had to pay demurrage on the *Cutty Sark* (which was likely to amount to more than £800) and that the owner of the *Linn Fern* was to claim more than £500.

In the book of cuttings is a preliminary notice of the sale by auction—by P. Burn & Son, auctioneers, 9 Exchange Place, Glasgow—of shipbuilders' plant, machinery, etc., at Scott & Linton's yard, Dumbarton, on Tuesday, 12th April 1870, 'a great portion of it nearly new and by first class makers'. Linton in his letter (30th April 1870) to Scott refers to the sale and gives the prices for various items to prove his point that some of the things in the yard were sold for almost nothing. A payment of two shillings in the pound was to be made to the creditors, but with the suggestion of an inquiry because of the 'fearful plunder' in the end. 'Your chest with some of my tools in it was sold for £2 but a lot of your tools and all my carvers' gouges are gone. . . . All the Models, Drawings and Paper . . . gone no person knows where.'

The business failure was not only the cause of Linton's serious financial difficulties, but also a personal tragedy. A son had been born on 23rd October 1869 at 24 High Street, Dumbarton, but for a time he and his family had to stay with relations since he was required to hand over the key of his house to Matthew Paul. (His furniture had all gone, for the benefit of creditors, before the end of January 1870.) About 10th December he had taken a post in Aberdeen in charge of the modelling and design department of Messrs Leckie, Wood & Munro, shipbuilders and engineers, at a fixed salary, with a percentage on the ships, and the prospect of a partnership later, on his discharge from bankruptcy. He was unable to get either a house or lodgings in Aberdeen and for some weeks his wife and family remained in Dumbarton. He brought them all from Dumbarton to stay with him in Aberdeen before the end of January 1870. He was offered a partnership in another concern to be opened in Aberdeen, and approached to enter a similar concern in the south. Nothing could be decided however until he had secured his discharge, and he was moved to press forward on this. At the beginning of April he had given up his place with Leckie, Wood & Munro and was hoping to visit Hull, London and Liverpool to secure orders and enter a new yard on 15th May. The name of the firm was given in his letter of 30th April as Morton, Wyld & Co., although his own name could not yet appear, and his letter (13th May 1870) to Scott from the Woodyard, Dumbarton, gives the impression that he was about to reopen the old Dumbarton yard and in two or three weeks to have laid down the keels of two screw steamers. It was unfortunate that Linton should have been concerned with this new firm reopening the Woodyard, because they were in the bankruptcy court in November 1870, and the shipbuilding yard was closed again before the completion of the first vessel contracted for.

In the *Lennox Herald* of 28th May 1870, reference is made to the Woodyard, the shipbuilding yard formerly occupied by William Denny & Bros, more recently by Scott & Linton, and 'now by the firm of Morton, Wyld & Co.'. It was important for the burgh that the premises should be extensively and profitably used. The firm had contracted to build a river steamer (screw) of about 300 tons, with the latest improvements for the comfort of passengers and a high rate of speed.

According to that same weekly newspaper of 17th September 1870, Morton, Wyld & Co. were said to have contracted with a Liverpool house to build a screw steamer of 1,000 tons. In the paper's annual review of shipping (issue of 31st December 1870) the annual tonnage of vessels contracted for was said to be 'exclusive of the river steamer and of the unfinished screw steamer on the stocks in the Woodyard'.

In the *Edinburgh Gazette* of 4th November 1870, notice was given of the sequestration of the 'estates' of Morton, Wyld & Co. on 2nd November 1870 and of a meeting to elect a trustee and commissioners on 11th November. The bankrupts were examined in the Sheriff Court of Lanarkshire on 1st, 13th and 23rd December and the proceedings were reported in the local newspapers, together with correspondence from interested parties. It would be interesting but a little unprofitable to follow the history of this small shipbuilding venture if it were not for Linton's concern with it. Briefly, the firm was constituted on 15th May 1870, with a capital of £1,500 (£1,000 from Morton and £500 from Wyld), and they took a lease of the Woodyard for ten years, with the option of a break at three. Morton came into shipbuilding after some concern with the boot and shoe trade in Dumbarton and two previous business failures. Wyld was responsible for the counting house. Morton claimed to have put about £1,150 into the business, including purchases made by him of Scott & Linton's effects and sums put out in wages to the men fitting up the articles bought. Wyld claimed that he had seen only one voucher (which was not handed over to him) for part of the machinery purchased—one by P. Burn & Son for £480. 19*s.* 1*d.*—and did not know whether or not the remainder had been paid by Morton. At his examination on 1st December 1870, Morton stated that the firm had entered into a contract for a screw steamer of 251 tons for Mr Barrow Moss, Liverpool. 'The contract was negotiated by myself and Mr. Linton on the one hand, Mr. Moss on the other, with of course the advice of Mr. Wyld. The hull of the steamer is pretty near completed, and could be launched in a few days.' There is no other reference to Linton in the proceedings, to judge from the newspaper reports.

Morton, Wyld's bankruptcy proceeded on official lines from the beginning; Scott & Linton's bankruptcy seems to have been dealt with 'unofficially' in the beginning, but official procedures had to be used when the two partners sought their discharge from bankruptcy. Their 'arrangement' with their creditors in September 1869, even although it was followed by the sequestration of their property, apparently had not the force of official bankruptcy proceedings.

Among the McCall letters is one from a Glasgow accountant, D. G. Hoey, dated 11th February 1870, to Scott, on the subject of Trust affairs, in which he states that half-a-crown in the pound may possibly be paid and concludes, 'so much for the success of amateur winding up!' On 21st July 1870 Scott and Linton jointly signed a brief formal letter to their lawyers applying to have the firm put into voluntary liquidation.

In the *Edinburgh Gazette* of 29th November 1870 (p. 1485) was published a notice that the estates of Scott & Linton were sequestrated on 28th November and the sequestration remitted to the Sheriff Court of Dunbartonshire. A meeting to elect the trustee and commissioners was held in Dumbarton on 9th December when David George Hoey (the Glasgow accountant) was appointed trustee, and Colin Campbell and John Scott junior (both iron merchants, Glasgow) and William Denny (shipbuilder, Dumbarton) were elected commissioners. The bankrupts were examined in the Sheriff Court House, Dumbarton, on Monday, 16th January 1871. D. G. Hoey postponed three times any declaration of dividend, until in February 1872 he announced that there were no divisible funds and that the commissioners authorised him to apply for his discharge. The last reference in the *Gazette* to this long drawn out business is in the issue of 6th December 1872 (p. 869) where a notice to the creditors of Scott & Linton refers to an order of the Lord Ordinary, dated 2nd December 1872, intimating that Scott & Linton had prayed to be finally discharged of all debts.

CHAPTER 5

In all likelihood, Hercules Linton and his family returned for a time to Glasgow and Dumbarton, after their stay of some months in Aberdeen. The Post Office Directory for Glasgow for 1872/3 mentions S. [*sic*] Linton, consulting shipbuilder and engineer, at 161 Hope Street (Lower 91 Pollok Street). A son, Alexander Linton, was born in Govan in January 1872, but died at 91 Pollok Street in October 1872, where another son, Alfred David, was born in August 1873.

Curiously, the Directory for 1873/4 gives for the same address, not Linton but W. Scott, Dundas & Co., consulting engineers. Scott's name was William Dundas Scott, but I cannot find anything about this company.

Hercules Linton's brief obituary notice mentions that when the firm of Scott & Linton went into liquidation, Linton went as assistant manager of Messrs Gourlay Bros' Camperdown yard in Dundee. His name appears in the Dundee Directory for 1874 and in the next issue of 1876, and in both entries he is described as 'shipbuilder' [15]. A daughter was born in Dundee in February 1876 and in November of that year he was elected a Fellow of the Society of Antiquaries of Scotland while still living there. His last letter to Scott, among the McCall documents, is dated 4th June 1877 from 2 Thornley Terrace, Dundee, and is an inquiry for a friend about the purchase of two miniature locomotive engines for passenger traffic.

Dundee Foundry was established about 1790 and by 1820 had turned to steam engines, acquiring national prestige by 1850. Gourlays, from Edinburgh, became part-owners of the foundry in 1846. In the 1860s the reputation of Gourlays as engineers and shipbuilders came from the ability of Henry Gourlay, who was a member of the Institution of Engineers & Shipbuilders in Scotland and of the Institution of Naval Architects. (He read a paper to the British Association in 1867.)

By 1854 Gourlays had become shipbuilding tenants of the Dundee Harbour Trust and occupied until 1870 a site at the east end of Marine Parade. At that time the traditional faith of shipowners and shipbuilders in wood was beginning to yield to iron, and from the beginning Gourlays concentrated on iron and on steam and screw. Although uninhibited by tradition, Gourlays made use of the pool of labour created by their predecessors and by 1866 they were the biggest of five effective shipbuilding firms in Dundee, employing about 300 men. The Harbour Trust had been reclaiming land down river, farther east from the existing shipyard area. This new ground—for the Camperdown yard—was offered by auction in December 1869 and let to Gourlays (the only bidder) to provide space for longer slipways.

There was a national boom in shipbuilding in the early 1870s following the opening of the Suez Canal and the expansion in world trade. At their Camperdown yard Gourlays continued their upward trend and in 1876, when Dundee's shipbuilding output reached a new level, Gourlays had seven of the twenty-three launches [16].

It is probable that Linton came to know Henry Gourlay during the time he was surveyor of iron ships for the Liverpool Underwriters' Association. And similarly during that time, from 1862 until 1864 and later, he came to know the Sunderland shipbuilder, T. R. Oswald, with whom he was associated in the 1880s and 1890s.

One of the few named and dated newspaper cuttings in the book of cuttings is from *The Highlander*, 17th June 1876, describing proceedings at the last meeting of the session of the Society of Antiquaries of Scotland, held at the Royal Institution, Edinburgh 'on Monday'. Mr H. Linton of Dundee gave a paper on the results of investigations recently made by him among the remains of the early settlements in the Sandhills of Culbin, near Findhorn, Morayshire. He found evidence of the manufacture in a bronze-like metal of such articles as needles, bodkins, hooks, and mixed with them, flint arrowheads, flint chips and flakes, with fragments of medieval pottery, iron knives, fish hooks and so on. A large and interesting collection of the objects found was exhibited.

In a letter dated 30th November 1876, from the secretaries (John Stuart and Arthur Mitchell) and the vice-president (John Alexander Smith) of the Society of Antiquaries, it was announced to Hercules Linton that at a general meeting of the Society held that day he had been elected a Fellow. This election was noted in a Dundee paper under the heading 'Honour to Townsman', where it was stated that 'Mr. Hercules Linton, of the Camperdown Shipyard [Messrs Gourlay Bros], has been elected . . . on the recommendation of Lord Kinnaird and other Fellows . . . in connection with his interesting discoveries among the sandhills of Culbin.'

An interesting short newspaper cutting, mounted in the old book of cuttings, refers to Linton's Dundee period. A sailing yacht, named *The Swift*, was lying in King William's Dock, Dundee. Of 12 tons register and flying the flag of the Glasgow Yacht Club, the vessel was being refitted and the understanding was that the owner—Sir James Linton-Foulis—was to attempt a voyage across the Atlantic with her. This newspaper paragraph was crossed out and a manuscript note in ink states that *The Swift* was built by H. Linton for his own use, 1876.

Linton is next to be traced in Glasgow, first in March 1878 at a Pollok Street address (district of Kinning Park) where a son was born, and two years later at Dennistown where a daughter was born. It is probable that he returned to Glasgow from Dundee to resume work as a consulting shipbuilder and engineer. Nothing definite is known of his work at this time, but in the book of cuttings is an extract relating to a shipbuilding yard in a rather out-of-the-way place in Kinning Park—a plot of ground on the corner of Great Wellington and Park Streets. There Linton was to construct a wooden screw steamer—80 feet keel, 10 feet beam, 7 feet deep, about 50 tons BM—to the order of a gentleman in Glasgow. This was said to be an experimental model very different from other steam vessels in that it would have fine lines for speed. Although there might be some difficulty in getting the ship to the Clyde, because of the distance, it was hoped to use steam power. After a few weeks it was hoped to lay down the keel of another steam vessel on similar lines, but to be fitted with Captain J. W. Sadler's patent double rudder and movable propeller, which was said to have many advantages over the present build of steamers.

Another mounted newspaper cutting, describing the model of a very experimental steamship, is also probably connected with Linton's Glasgow period after his stay in Dundee. This very remarkable screw steamer was to be built on the Clyde for the Compagnie de Navigation Economique (Limited) of Glasgow and Paris, and the finished large-scale model (about 13 feet long) was on view at the office of Messrs Boustead & Dunbar, of Dixon Street, Glasgow.

This vessel will have no funnel of any kind, burning her own smoke, and her four masts will be placed in pairs on the gunwale on projecting brackets to keep them clear of the rails, boats, davits, etc. when folded down, this being done when the vessel is going head to wind, so as to lessen the resistance offered by the masts, spars, sails, etc. On each mast will be placed a large boom and gaff, and between the masts will be placed a large square sail and topsail running on jackstays. The total width of sail set on each pair of masts when running before the wind will be 147 feet in width and 127 feet in height. The effective horse-power will be 1500, so as to give the vessel a high rate of speed, and the fuel required will be only one-half that of an ordinary compound marine engine. The result is obtained by superheating the water in the boiler by the introduction of a substance which need not be named, and which substance is variable according to circumstances. The space occupied by the boilers, engines, and coal bunkers will only be one-half that of an ordinary steamer of the same power. By this arrangement the steam pipes, etc. will be only about one-fourth the size of those of an ordinary steamer. Mechanical ventilation is to be substituted for the present funnel system, because it is possible to thoroughly exhaust the heat and burn all gases. Instead therefore of a thick black smoke issuing from the funnel, with burnt air at a temperature of 500 to 600 degrees Fahrenheit, it will be drawn off by the aid of appliances connected with the new system at a temperature of 150 to 175 degrees, and this of itself is no mean result; besides, owing to the change in the boilers, the present system of firing will be entirely superseded, and an improvement introduced by which the overpowering fatigue endured by stokers in hot latitudes will be wholly done away with. The engines, boilers, masts, etc. are the invention of Charles Tellier, Esq., of Paris, engineer. The model and lines were designed by Mr. Hercules

Linton, marine architect, of this city, and are a modification of the lines of the celebrated China clipper *Cutty Sark*, built by the same gentleman. The vessel is to be constructed for great speed, having very fine lines and a good rise of floor. Her extreme dimensions are—length 285 feet, beam 35 feet, depth 24 feet. As she is intended for long voyages, the screw is to be constructed on a new plan, so that it may be raised out of the water when the vessel is under sail. To show fully the working of the new arrangement of masts and sails, a full model of large size (nearly thirteen feet long) has been made, with all the appliances for lifting and lowering the masts, and showing the new disposition of the rigging, etc. The deckhouses are constructed so as to offer as little resistance as possible to the wind. The cabins are to be fitted up handsomely, and will be slightly different from ordinary steamers, being mostly of an octagonal form, with the angles relieved by mirrors, etc. The vessel is to be immediately proceeded with, after it has been determined whether she is to be built of steel or iron. On completion of the hull, the engines and boilers will be brought from Havre, where they are being made, after which she will proceed on her trial trip a long distance seaward, with a select party of scientific and nautical men on board, to test the economy of the new system.

It is not known, and it may be doubted, whether such a steamship as modelled was ever built.

CHAPTER 6

There is a tradition in the family of contact between Hercules Linton and T. R. Oswald, the Sunderland shipbuilder.

Thomas Ridley Oswald had an amazing career. He was four years younger than Hercules Linton, but had opened a yard at Pallion, on the south bank of the Wear, known as High Yard, in 1859 when he was nineteen. The site was owned by Christopher Maling Webster, whose daughter Oswald married, and it is very probable that the young shipbuilder obtained financial help from his father-in-law. Very soon he opened a second yard at the North Dock, and as he began as a builder of iron ships, his premises had to be equipped with plant and machinery far in advance of that required for wood shipbuilding.

Eight barques and full-rigged ships were built at Pallion during the first year, each of them about 500 tons. In 1861 the 898-ton *Ironside* was launched for Temperley, Carter & Darke, of London, and in that same year Oswald encountered the first of his financial difficulties. Within a few months of his first failure he was again building iron sailing ships for noted owners in London and Liverpool, but he had given up his North Dock yard. His first steamer, the *St Thomas* (brig-rigged, 1,274 gross tonnage) was launched in 1863, but splendid barques and full-rigged ships came from the Pallion berths in 1862–4. Nine of these:

1862 *Acapulco*—ship 898 tons
Chanaral—barque 589 tons

1863 *Thomas S. Stowe*—barque 498 tons
Jessie Jamison—barque 504 tons
Saltara—ship 939 tons
Dolbadern Castle—ship 989 tons

1864 *Corrientes*—barque 598 tons
Carrizal—barque 598 tons
Otter—barque 499 tons

are named in Linton's list of ships surveyed by him, no doubt when he was with the Liverpool Underwriters' Association.

Top-class sailing ships and more and more steamers were launched in the months that followed and Oswald was quick to take advantage of the trend in the late 1860s to build iron clippers designed for the speedy transport of emigrants and goods to Australia and New Zealand. The result was the building of ten outstanding ships which included the *Cashmere* (1868, 1,245 tons), *Hoogly* (1868, 1,300 tons) and *Penthesilia* (1869, 1,700 tons).

Not content with one of the largest yards on Wearside, Oswald set out to build an industrial empire. At Castletown, on the north side of the river, he built his own ironworks and rolling mills, and imported iron ore from Spain for the plates from which his ships were fashioned. He built the engines and boilers to power those ships. His hundredth ship, the *Humboldt* (about 1,900 tons), was launched in 1871 and in that year his engine works were completed. Although 1872 began well with the building of three stately ships, Oswald failed for £300,000. The Castletown works and Pallion High Yard were closed and Oswald's father-in-law, having lost faith in his son-in-law's financial if not shipbuilding abilities, advertised the lease of the Pallion yard. But in spite of this financial disaster, Oswald was soon formulating plans and within twelve months had taken over two old yards on the Pallion side and was launching again, iron clippers for London and Liverpool owners. In 1875 he launched the clipper ship *Sierra Morena* for the famous Sierra Line owned by Thompson & Anderson, the last ship he built in Sunderland. After barely two years and the launching of about eleven large vessels, his creditors were again pressing him and this time his career on Wearside was ended. Over a period of sixteen years, 149 ships had been built, but he continued to build ships for another twenty years [17].

The Pallion yard was dismantled and shipped to Woolston, near Southampton, where a new yard was opened in 1876. The first vessel to be laid down was the barque *Aberfoyle* (853 gross tons), which was given the yard number 150. She had a flying start as her frames were brought from Sunderland and already bent to shape. In 1877 the title of the firm was changed to Oswald Mordaunt & Co., and a

reputation was soon established for building large iron sailing ships which were efficient dead weight carriers, the only type that could still earn a profit in the closing years of sail. Many of these ships were astonishingly long lived and several were converted to oil or molasses carriers. Thus the *Monterey* built as No. 172 in 1878 became the tanker *Cypromene* and was only sold to the shipbreakers in 1934, after weathering the storms for 56 years.

The sail plans of the Oswald Mordaunt vessels included ship rig, barques, brigs, brigantines, schooners and ketches. No. 200 the *Bitterne* was one of the few full power ships that were built by the firm and was also the largest. Her length was 382 feet, the tonnage 5,085, the loaded displacement 9,250 tons, and the speed of 13.3 knots was due to machinery that developed 2,266 i.h.p. on trial. In her day she was reported to be the largest cargo ship to enter the Port of London. No. 253, the *Ville de Dieppe*, was the last ship to be built before Oswald, that somewhat peripatetic shipbuilder, had moved on to Milford Haven. No. 253 was a steel and iron barque carrying petroleum in bulk. She was delivered in 1889, so that in the short space of thirteen years the firm had turned out some 104 ships mostly of large size for the period [18].

After his fourth financial failure, this time at the Woolston yard, Oswald moved to Milford Haven where he opened a small yard which was still operating in the 1890s. Towards the end of the nineteenth century, Milford Haven had recovered much of its earlier prosperity and had become the headquarters of a steam trawling fleet and a fish-curing station. Oswald is said to have built there a number of steamers and trawlers and to have been in very reduced circumstances. Long afterwards in Sunderland he was referred to as a 'fine old gentleman'.

Linton went to Woolston in 1880 and apparently stayed there until 1884. He is also said to have joined Oswald in Milford Haven for two years in the early 1890s. It is not known whether he was concerned with the design of the Oswald Mordaunt ships; he may have acted as yard manager. It is tempting to think of his being concerned with the design and building of pioneer oil tankers. (A daughter was born in Woolston in October 1882 and in that connection he described himself as 'marine architect'.)

CHAPTER 7

The remaining years in the life of Hercules Linton can be briefly described.

He returned north from Southampton, this time to Montrose where his last child, a daughter, was born in December 1884. His wife died shortly afterwards on 30th January 1885, a sad blow as he said, which left him 'like a rudderless ship', with a family of 'motherless nine', while his Misey (Marjory) had joined their tenth in heaven.

Heading to an obituary notice of Hercules Linton in May 1900.

In directories until 1887 he is to be traced at two addresses in Montrose, as marine architect, but there are no other details of his work.

His bachelor uncle, William Anderson of Bervie, died in 1888 and his estate passed to Hercules and his sisters. It is presumed that shortly after his uncle's death he took his family to live in Bervie, where the youngest children were looked after by relations. He may have left his family for about two years when he went to Milford Haven; but he seems to have interested himself in antiquarian matters and more and more in local affairs. As might be expected, he still maintained an interest in seafaring concerns. In 1896, with the help of one or more of his sons, he built a launch for his friend James Buchanan Gibb, and he may have joined him in a small coastal trading venture. And when a Danish schooner, *Anna Maria*, bound for Aberdeen with a cargo of Norwegian granite, was wrecked near Gourdon in the early hours of Saturday, 30th November 1895, and became a total wreck, at the salvage sale the following Wednesday Linton purchased 140 tons of granite for £74. The granite, or what remains of it, has not yet been recovered!

He was elected to the Town Council of Inverbervie (Bervie) in November 1895 and attended his first council meeting on 27th December in that year. He remained a councillor until his death, and from the minutes it appears that the last council meeting he attended was held in March 1900 [19]. According to a private letter, the weather in January and February 1900 was frightfully cold, with the mustard freezing in the cupboard and the soapy water in the kitchen. In spite of a chill and sciatica, he was making fishing rods and looking forward to visiting Edinburgh for the convention of Royal Burghs on the first Tuesday in April, with a grand dinner at night. He had been appointed by the Town Council to go as Bervie's representative and there is no reason to think that he did not make the journey. He died suddenly of heart trouble on 15th May and was buried on 19th May 1900. The brief obituary notice already referred to mentioned his great interest in the affairs of the Royal Burgh of Inverbervie and the high esteem in which he was held for his kindly disposition.

NOTES

1 See George F. Black, *The Surnames of Scotland*, New York, 1946, p. 430; and George Hay, *History of Arbroath*, 2nd ed., 1899, p. 70.

2 *The Statistical Account of Kincardineshire by The Ministers of the respective Parishes*, 1843. Parish of Bervie by the Rev. John Glegg, Minister, p. 11.

3 Peter F. Anson, *Fishing Boats and Fishing Folk on the East Coast of Scotland*, 1930, p. 223.

4 I am indebted to the Secretary of Lloyd's Register of Shipping (and especially to Mr G. W. Drucquer) for this and other information. See also *Annals of Lloyd's Register*, 1934, and George Blake, *Lloyd's Register of Shipping, 1760–1960*, 1960.

5 See *Arbuthnott; with some Reminiscences of the Parish sixty years ago.* By Geo. Clark Suttie, F.S.A. SCOT., 1907. Montrose, 1908.

6 Arbroath High School is the direct descendant of the old Academy and I am indebted to the Rector, Mr R. G. Hay, for replying to my inquiry.

7 The indenture cannot now be found, but was produced by his son, the late Mr H. A. Linton, at an interview with a newspaper reporter on 13th March 1928, and referred to in the Aberdeen *Press and Journal* of 14th March 1928. No reference was made to the terms of the apprenticeship except to the one condition of avoiding 'all immoral practices, especially that of using intoxicating liquors and tobacco'. The usual length of apprenticeship at that time was six years, and Linton would have completed it at the end of December 1860.

In September 1966 I was informed by Mr A. E. Gordon of Hall, Russell & Co. Ltd that Hall's had a serious fire in their office about 1946 and that no records are now available for the period 1855–60. There is in existence, however, an early photograph of Hall's

Shipbuilding Yard, 1862, with workmen and office staff against a background of two ships on the stocks. The photograph was reproduced in the Aberdeen *Press and Journal* of 21st December 1963, in an article on Footdee, the shipbuilding district of Aberdeen, by Cuthbert Graham and Ian Hardie. (I am very grateful to Mr G. Ross of Aberdeen Journals Ltd for tracing this reproduction of the 1862 photograph.)

I am also indebted to Mr D. I. Moor (Vickers-Armstrongs Ship Model Experiment Tank, St Albans) for drawing my attention to two earlier reproductions of this old photograph, now used again in connection with Hercules Linton. The first was in an article by Boyd Cable, 'The World's First Clipper', in *The Mariner's Mirror* (vol. 29, no. 2, April 1943, pp. 66–91). Boyd Cable gave much interesting information about Alexander Hall, the founder of the firm, and his large family of five sons and eight daughters. He discussed the experiments carried out before the schooner *Scottish Maid* was built in 1839, in which different shaped hulls were pulled by weights through water in a glass tank topped by red turpentine; and commended the public-spirited action of the firm in publishing a report on their experiments instead of patenting their 'Aberdeen bow'.

The second, and more important because more detailed, reproduction was in David R. MacGregor, *The China Bird* (The history of Captain Killick and one hundred years of sail and steam), 1961. Plate 13 (facing p. 94) identifies the two vessels on the stocks (left *The Coulna Kyle*, 579 tons; right *The Natal Star*, 336 tons) and every man and boy in the photograph. By 1860 all Alexander Hall's large family were dead except James and William, respectively second and fourth from the right in the back row. James was mainly the business head, William the designer and builder. The rest of the group includes three carpenters, five foremen carpenters, a boat builder, a shipbuilder, a foreman joiner, two blacksmiths, a blockmaker, a gatekeeper, a cashier, an office clerk and an office boy, as well as Alexander Hall, junior, and William Hall, junior.

The firm of Hall, Russell, now the main shipbuilding concern in Footdee, was formed in 1864 by the Halls, in association with Thomas Russell, an engineer, to build engines and boilers for their ships, but in 1867 it began to build ships on its own account. In recent years the firm has operated with Messrs Alexander Hall & Co. as a single unit.

8 See Victoria E. Clark, *The Port of Aberdeen* (A history of its trade & shipping from the twelfth century to the present day), Aberdeen, 1921. Messrs Alexander Hall & Co. were able to supply two illustrations for this book: the *Scottish Maid*, 1839; and the Waterbridge, Footdee.

In spite of Boyd Cable's claim, the *Scottish Maid* was not 'the world's first clipper', but the first vessel with the 'Aberdeen bow', in the tradition of Aberdeen fast sailing schooners. Apparently these schooners, enlarged and rigged as barques or ships, exerted a greater influence on the design and appearance of fast merchantmen than ships launched from any other single port. Such vessels were loosely termed 'clippers'. When the old navigation laws came to an end as from 1st September 1849, the new rules from the beginning of 1850 encouraged shipowners to order, and shipbuilders to build, vessels of a new design, and especially to enter into competition with American ships. (See D. R. MacGregor, *The Tea Clippers*, 1952.)

9 McCall documents, National Maritime Museum (MS.59/013). The letter of 17th March 1868 was quoted by Mr Frank G. G. Carr, Royal Institution of Naval Architects *Quarterly Transactions*, July 1966, p. 215.

At a meeting of the Royal Institution of Naval Architects on 24th March 1965, Mr Carr (then Director, National Maritime Museum, Greenwich) read a paper, 'The Restoration of the *Cutty Sark*', which was followed by an oral discussion. The paper and a report of the discussion, together with written contributions and a reply by Mr Carr, were printed in the *Quarterly Transactions* for July 1966.

One of those present at the meeting was Mr W. W. Scott-Moncrieff, a son of William (Dundas) Scott (11th May 1846–1st December 1924) who in his early years had followed his father in dropping the name Moncrieff. Mr W. W. Scott-Moncrieff gave some very interesting sidelights on his father's work and personality, and I am indebted to him and to his sister, Mrs M. McCall, for information about their father. In March 1959 Mrs McCall presented some thirty-five documents, relating to the Scott-Linton partnership, to the National Maritime Museum.

William (Dundas) Scott-Moncrieff was educated at Jedburgh and Edinburgh Academies and served his time as an engineering apprentice in a Clyde shipyard. He is said to have won medals (bronze and silver), and a silver tankard for inventing a 'governor' for a ship. It is probable that Linton came to know him during his apprenticeship. It is said that Scott in later years never referred to the *Cutty Sark* period of his life or spoke about it to his family, and that before his death in 1924, he was quite surprised at the discussions on the preservation and restoration of the 'clipper' ship. He was a very versatile man with very definite literary gifts. He invented a tramcar propelled by compressed air, which was built and ran in Glasgow (1877); he wrote fourteen of the twenty-seven chapters on 'Iron and Steel' in the series, *Great Industries of Great Britain* (1877 *et sqq.*); and, from about 1885 until 1907, he registered numerous patents in connection with sanitary fittings and the disposal and treatment of sewage. He was especially concerned with the layout and arrangement of filter beds, for the bacterial treatment of sewage from large country houses, barracks, hospitals, and part of the town of Towcester. In May 1908 he gave the Chadwick Lectures (later printed) in the University of London on 'The Engineering Aspects of Recent Advances in connection with Sewering'.

By 1872 he had written (and had published anonymously) a verse drama in five acts on Mary, Queen of Scots. This he rewrote in three acts and had published in 1881, in folio, with seven illustrations drawn and etched by distinguished artists of the period. It was dedicated to Emily Mary de Gilibert—the lady whom he married later that year—'in the hope that at some future time I may draw a portrait more worthy of her sex, and a character as charming as her own'. They had three sons and one daughter. Scott also published two volumes of poems: *Amor Amoris: Songs, Sonnets and Lyrics* (1901, 3rd ed., 1924), and (posthumously) *Wind Songs: being later poems* (1925).

10 I am indebted to Mr W. M. Martin, Dumbarton Burgh Librarian, and his staff for help in consulting the library's copies of these early newspapers and for answering inquiries.

In 1922, Captain George Moodie, the first captain of the *Cutty Sark* and the representative of the owners who supervised the building of the vessel, presented to the library his working copy of the agreement for the building of the ship, and his three log books, 1870–2. I have to thank the Dumbarton Library Committee for permission to have photostats of the agreement and of certain entries in the log books.

11 For each such vessel a file was gradually built up, from her First Entry Report—on her completion, after her construction had been watched and examined by Lloyd's surveyors while she was on the stocks and after launching—to her last survey, with repair or modification to warrant her continued inclusion in the Register. These early reports are at the National Maritime Museum, but not those of the *Cutty Sark*. The original First Entry Report on the *Cutty Sark* was presented by the Chairman of Lloyd's Register to the *Cutty Sark* (Preservation) Society on 30th April 1953. A photographic copy is retained by the Museum.

12 *R.I.N.A. Quarterly Transactions*, July 1966, p. 210 (contribution by the late Mr J. M. Murray); pp. 213 and 214 (contribution by Mr D. I. Moor).

A careful reading of the correspondence attached to the First Entry Report on the *Cutty Sark* gives a slightly different interpretation.

See also Basil Lubbock, *The Log of the Cutty Sark*, 2nd ed., 1945, chap. II, especially pp. 20–7 and 43–6.

13 Captain Moodie's working copy of the agreement is now in the Dumbarton Public Library (see note 10).

See also C. Nepean Longridge, *The Cutty Sark*, 1933—a description of the hull, deck fittings and rigging, with a detailed account of the building of a scale model—in two volumes: Vol. I, introd. p. xxvi. In Vol. II, App. 1 (pp. 205–16), is printed the original specification for the building of the *Cutty Sark*, kindly supplied to Dr Longridge by Captain and Mrs Dowman. (It is said—Carr, 'The Restoration of the *Cutty Sark*', p. 203—on the authority of Scott's son, Mr W. W.

Scott-Moncrieff, that the specification was discovered by Scott in an old tin deed-box in the loft of his house about 1923/4, when Captain Dowman was re-rigging the *Cutty Sark*, and made available to him.) The ship may not have been finished in exact accordance with the specification, which is amazingly detailed—hull, keel, frame, floors, keelson, bilge plates, stringers, diagonal ties, beams, stanchions, bulkheads, planking, rails, steering gear, hatches, brasswork, rigging, masts, yards, painting, cementing, and furniture and equipment, including a double suit of sails. The contents of the stores for the carpenters and boatswain, the coopers and shipchandlers (including mops, brushes, fishing lines and signal-flags), and the cook, are listed in great detail. Cabin stores were to include among much else a copper tea kettle, knives, forks and spoons in Elkington electro-plate, certain muskets with bayonets, gunpowder and cartridges, brass candlesticks with snuffers, and hen coops made of teak!

Incidentally, Dr Longridge (Vol. I. p. xxiv), on the evidence of Captain Millett, who served his time on both the *Tweed* and the *Cutty Sark*, discounts the suggestion that Linton used the lines of the *Tweed* for the *Cutty Sark*. Captain Millett described the *Cutty Sark* as an uncomfortably wet ship because of her extremely sharp entrance.

14 I am greatly indebted to Dr P. L. Payne, Colquhoun Reader in Business History, University of Glasgow, for help in making it possible for me to consult the Denny records.

In *Denny Bros Dumbarton, 1844–1932*, a lavishly illustrated souvenir volume published in 1933 for William Denny & Bros Ltd (for private circulation), there is much interesting information about the personnel of the firm and the ramifications of their business partnerships. For instance (p. 10) the Woodyard was first leased from Denny and Rankin in 1845 and fitted with plant for iron shipbuilding. The yard was bought outright by William Denny (the second) in 1847. In 1849 the firm of Denny Bros was dissolved by mutual consent and a new firm was formed under the name William Denny & Bros. William Denny (the second) died in June 1854, his brother James (the first) retired in 1862, and thereafter Peter (the second) was left as the sole surviving partner of William Denny & Bros (pp. 12–13).

The conditions of tenure of the Woodyard, now the responsibility of the trustees of the late William Denny (the second), were felt to be unsatisfactory and in 1864 a piece of ground was taken on the other (east) side of the river, which became the nucleus of the later shipyard. When the lease of the Woodyard expired in 1867, the whole concern was moved to the new yard named the Leven Shipyard.

The *Cutty Sark*, although not built by Denny, was included in the souvenir volume among outstanding vessels built by them over the years, and various facts were given about her. She was laid down in 1869 at 'the Sandpoint, just opposite Denny's new yard. . . . Her contract price was £17 per ton with a specified maximum of 950 tons. This was probably far too low a figure, especially in view of the extremely high standard of material and workmanship which was exacted by the owner; at any rate, the builders went bankrupt shortly before the hull was ready for launching. The creditors arranged for William Denny & Bros to complete the ship, which they did, bringing her over to their own dock in the Leven Shipyard after the launch. . . . In sailing weather the only ship which approached her in speed was her famous rival the *Thermopylae*. . . . It is interesting to note that, at 20′ draught and at 16½ knots, the E.H.P. of the *Cutty Sark* was of the order of 2250, so that if she had been power-driven her machinery would have been required to develop nearly 4000 I.H.P.'

Peter Denny (the second) was to guide the fortunes of the firm for 51 years. He died on 22nd August 1895 in his seventy-fourth year. His eldest son, William (the third), was apprenticed to Scott & Linton. Among the McCall papers (see note 9), in the draft or copy of a letter to his father of 17th August 1868, Scott mentioned that young Denny's apprenticeship was expected in a short time, but the premium of £300 (i.e. £50 per aunnm for six years) was not counted on. Among those same documents, in the statement of the accountants (McFarlane & Hutton of Glasgow) dated 17th September 1869, there is an item—W. Denny's apprenticeship fee £50—mentioned under 15th May 1869. This William Denny was twenty-one on 25th May 1868. (See Alexander Balmain Bruce, D.D., *The Life of William Denny, Shipbuilder, Dumbarton*, 1889.) He was admitted to partnership in the firm in 1871. In December 1870 he had been appointed one of three commissioners under a trustee to inquire into the business affairs of Scott & Linton.

15 I am indebted to Mr W. S. Taylor, Dundee City Librarian, for this information and other help.

16 See S. G. E. Lythe, *Gourlays of Dundee: the Rise and Fall of a Scottish Shipbuilding Firm*, Dundee, 1964, Abertay Historical Society Publications, No. 10.

17 For information about the career of T. R. Oswald I am indebted to Mr J. T. Shaw, Director of Libraries, Sunderland, who directed me to two articles by 'Blue Peter' in the *Sunderland Echo* (6th & 13th August 1964), and an entry in 'Wear Shipbuilders', a two-volume manuscript work by J. W. Corder, a local historian.

18 See K. C. Barnaby, *100 Years of Specialised Shipbuilding*, 1964, p. 49.

The late Mr Barnaby wrote the book for the centenary of the firm of Messrs John I. Thornycroft & Co. (Woolston Works, Southampton) who bought the yard in 1904. I am much indebted to him for information in private correspondence.

19 I am indebted to the present Town Clerk of Inverbervie, Mr W. B. Agnew, and his staff for this information.